DEVELOPING MATHEMATICS

100% NEW

Photocopiable teaching resources for mathematics

COUNTING AND UNDERSTANDING NUMBER

Ages 4–5

Hilary Koll & Steve Mills

A & C Black • London

Contents

Use language such as 'more' or 'less' to compare two numbers

Use ordinal numbers in different contexts

Recognise numerals 1 to 9

Published 2008 by A & C Black Publishers Limited
38 Soho Square, London W1D 3HB
www.acblack.com

ISBN 978-0-7136-8436-0

Copyright text © Hilary Koll and Steve Mills 2008
Copyright illustrations © John Haslam 2008
Copyright cover illustration © Piers Baker 2008
Editors: Lynne Williamson and Marie Lister
Designed by HL Studios, Oxford

The author and publishers would like to thank Catherine Yemm and Corinne McCrum for their advice in producing this series of books.

A CIP catalogue record for this book is available from the British Library.

Printed and bound in Great Britain by Martins the Printers, Berwick-on-Tweed.

A & C Black uses paper produced with elemental chlorine-free pulp, harvested from managed sustainable forests.

Introduction

100% New Developing Mathematics: Counting and Understanding Number is a series of seven photocopiable activity books for children aged 4 to 11, designed to be used during the daily maths lesson. The books focus on the skills and concepts for Counting and Understanding Number as outlined in the Primary National Strategy *Primary Framework for literacy and mathematics*. The activities are intended to be used in the time allocated to pupil activities in the daily maths lesson. They aim to reinforce the knowledge and develop the skills and understanding explored during the main part of the lesson, and to provide practice and consolidation of the learning objectives contained in the Framework document.

Counting and Understanding Number

This strand of the *Primary Framework for mathematics* is concerned with helping pupils to develop an understanding of the relationships between numbers and the way our number system works. It includes all aspects of counting, ordering, estimating and place value, and involves building awareness of how numbers can form sequences and can be represented on number lines and in grids. Also included in this strand of the curriculum is work on negative numbers, fractions, decimals, percentages, and ratio and proportion. Broadly speaking, this strand addresses topic areas that were described under the 'Numbers and the Number System' strand title of the former National Numeracy Strategy *Framework for teaching mathematics*.

Counting and Understanding Number Ages 4–5 supports the teaching of mathematics by providing a series of activities to develop essential skills in counting and recognising numbers. The following learning objectives are covered:

*   say and use number names in order in familiar contexts;
*   know that numbers identify how many objects are in a set;
*   count reliably up to 10 everyday objects;
*   estimate how many objects they can see and check by counting;
*   count aloud in ones, twos, fives or tens;
*   use language such as 'more' or 'less' to compare two numbers;
*   use ordinal numbers in different contexts;
*   recognise numerals 1 to 9.

Extension

Many of the activity sheets end with a challenge (**Now try this!**) which reinforces and extends children's learning, and provides the teacher with an opportunity for assessment. These might include harder questions, with numbers from a higher range, than those in the main part of the activity sheet. Some challenges are open-ended questions and provide opportunity for children to think mathematically for themselves. Occasionally the challenge will require additional paper or that the children write on the reverse of the sheet itself. Many of the activites encourage children to generate their own questions or puzzles for a partner to solve.

Organisation

Very little equipment is needed, but it will be useful to have available: coloured pencils, counters, dice and spinners, cubes, scissors, glue, squared paper, number lines, number grids and number tracks.

Where possible, children's work should be supported by ICT equipment, such as number lines and number tracks on interactive whiteboards, or computer software for comparing and ordering numbers. It is also vital that children's experiences are introduced in real-life contexts and through practical activities. The teachers' notes at the foot of each page and the more detailed notes on pages 6 to 13 suggest ways in which this can be done effectively.

To help teachers select appropriate learning experiences for the children, the activities are grouped into sections within the book. However, the activities are not expected to be used in this order unless stated otherwise. The sheets are intended to support, rather than direct, the teacher's planning.

Some activities can be made easier or more challenging by masking or substituting numbers. You may wish to re-use pages by copying them onto card and laminating them.

Accompanying CD

The enclosed CD-ROM contains electronic versions of all the activity sheets in the book for printing, editing, saving or display on an interactive whiteboard. Our unique browser-based interface makes it easy to select pages and to modify them to suit individual pupils' needs. See page 14 for further details.

Teachers' notes

Brief notes are provided at the foot of each page, giving ideas and suggestions for maximising the effectiveness of the activity sheets. These can be masked before copying.

Solutions and further explanations of the activities can be found on pages 6 to 13, together with examples of questions that you can ask.

Whole class warm-up activities

The tools provided in A & C Black's *Maths Skills and Practice* CD-ROMS can be used as introductory activities for use with the whole class. In the *Maths Skills and Practice* CD-ROM R, the following activities and games could be used to introduce or reinforce 'Counting and Understanding Number' learning objectives:

- *How many?*
- *Martians*
- *Racing*

The following activities provide some practical ideas which can be used to introduce or reinforce the main teaching part of the lesson, or provide an interesting basis for discussion.

Making mistakes

This activity focuses on counting in order. Use a hand puppet to 'say' a sequence of number names. Tell the children that the puppet sometimes makes mistakes and they should wave their hand if they spot one. Start counting from any small number and make a mistake, for example: *Six, seven, eight, ten, nine.* Include number name errors, for example: *Eleven, twelve, threeteen.* or *Eighteen, nineteen, tenteen.* Try choosing a larger number and counting backwards. Use the puppet to count forwards and backwards in twos, fives or tens.

Postman's journey

On a board, sketch a simple map showing houses and roads linking some of them. Write some numbers into each house.

Call out number names and ask the children to find the houses to show the journey made by a postman, for example 7, 4, 9, 3, 1, 5.

Hand out!

Ask the children to sit in pairs. Player A closes his/her eyes and holds out one hand, palm upwards. Very slowly, player B traces a number with their finger onto the palm (0, 1, 2, 3, 4, 5, 6...). Player A must guess what number has been traced.

'I spy three of something. What could they be?'

Play this game as a whole class. Look for items that are the same, such as six books, four signs, three doors, two teachers and so on.

Numbers around us

Discuss with the children where they see numbers in the environment: in the home, the street, on TV, in shops and so on. Talk about how numbers are ordered or arranged, for example the house numbers on one side of a street going up in twos, a microwave oven timer counting back in ones. Also discuss how numbers are used as labels (cardinal numbers) such as numbers on a bus, page numbers, TV channels, car registrations and so on.

Notes on the activities

Say and use number names in order in familiar contexts

Hearing and using numbers in a range of familiar contexts, such as rhymes and stories, is an important element of early number work. The following rhymes can be used in addition to the ones on these pages:

One, two, buckle my shoe...
One, two, three, four, five, once I caught a fish alive...
One potato, two potato...
One man went to mow...
There were ten in a bed and the little one said...
One for sorrow, two for joy...
Five little speckled frogs...
Ten green bottles...
Five currant buns in the baker's shop...
This old man, he played one...

One having fun (page 15)

Say the following rhyme several times and encourage the children to join in.

One having fun,
Two in a queue,
Three in the sea,
Four in a store,
Five by a hive.

Actions can be added or children asked to participate in acting this out, for example 15 children could perform the following actions while the others say the rhyme:

One having fun,	One child dances
Two in a queue,	Two stand in a line
Three in the sea,	Three pretend to swim
Four in a store,	Four hold pretend shopping baskets
Five by a hive.	Five wave their hands around, waving away imaginary bees

Ask the children to make up their own actions and rhymes to match the numbers on the sheet, for example: three buzzing like a bee.

SUGGESTED QUESTIONS:

- How many children are: in a store? by a hive? in the sea? etc.
- How many more children are by a hive than in the sea?
- Can you think of a rhyme for six? or seven?

One baby bunny (page 16)

The 'One baby bunny' rhyme is sung to the same tune as 'Ten green bottles', but each verse begins with one more baby bunny, for example the second verse is 'two baby bunnies'. This continues until there are five (or ten) baby bunnies playing in the sun.

Actions can be given to this rhyme using one hand. Ask the children to hold up one hand, palm forwards and wiggle one finger in the first verse, then wiggle two fingers in the next verse, then three, then four and so on.

SUGGESTED QUESTIONS:

- What is one more than five?
- How many bunnies were there at the end?

The zoo rhyme (page 17)

Say the following rhyme several times and encourage the children to join in.

One, two let's go to the zoo!
Three, four hear the tigers roar!
Five, six, watch the monkeys' tricks!
Seven, eight, feeding time is late!
Nine, ten, see the lions in their den!

SUGGESTED QUESTIONS:

- Which number rhymes with 'roar'? Show me four fingers.
- Which is the largest number in the rhyme?

One computer... (page 18)

A fun way to repeat this rhyme several times is to begin saying it very slowly and gradually speed up until it becomes very difficult to say it quickly (like a tongue-twister). Another way is to say each number of computers and ask the children to clap that many times, for example 'one computer' CLAP; 'two computers' CLAP, CLAP; 'three computers' CLAP, CLAP, CLAP, etc.

- Did you find it easy to put the cards in order?
- Have you arranged your cards from smallest to largest or from largest to smallest?
- (When holding two cards) This card shows one computer and this shows three. How many altogether?

Running track (page 19)

Children will need a cube each, some foil-wrapped counters and a spinner marked 1, 2, and 3 for this activity. It may be useful to photocopy this sheet onto A3 paper.

SUGGESTED QUESTION:

- Which of these numbers is the smallest? Which is the largest?

Know that numbers identify how many objects are in a set

Children gain an understanding that numbers identify how many objects are in a set through many experiences of sets of objects whose only commonality is number, for example, three sweets, three cats, three chairs, etc. Thus many fairy tales and stories involve sets of animals to help children understand cardinal numbers, such as for the number three: *Goldilocks and the three bears, The three little pigs, The three billy goats gruff,* etc.

Tea party (page 20)

Encourage the children to count the number of teabags in each teapot and record the numeral in order to help them recognise any sets with the same number.

SUGGESTED QUESTIONS:

- How many bags are in this teapot?
- Which teapot has the most/fewest bags?

Hello, campers! (page 21)

At the start of the lesson, invite the children to identify sets of objects that have the same number, such as three pencils, three chairs, three girls. Ensure that the children understand that the items do not need to be the same, it is the number of them that is the same.

SUGGESTED QUESTIONS:

- How many tents here?
- Which field has the most/fewest tents?

Frying fun! (page 22)

This activity could be a follow-on activity to this familiar rhyme:

Ten fat sausages sizzling in a pan
Ten fat sausages sizzling in a pan
One went pop!
and another went bang!
There were eight fat sausages....

SUGGESTED QUESTIONS:

- Which pan has the most/fewest eggs?
- How many more eggs are in this pan, than this one?

Magic beans (page 23)

The extension activity provides practice in drawing sets of objects to make equal-sized sets. Invite the children to collect counters or cubes so that they have the same number in each hand.

SUGGESTED QUESTIONS:

- Have you checked your answers by counting them again?
- How did you make sure you counted each bean once?
- How many more beans are in this hand than that hand?

Count reliably up to 10 everyday objects

Children need to be able to say the number names in order before accurate counting can begin. The next step for them is to point to objects systematically and one at a time as the number names are said. This action is known as one-to-one correspondence, where one name is matched to one object. Children also need to realise that the number they say as they point to the last object tells them how many items there are in a group. Eventually they will learn to recognise small sets of objects without counting them (sometimes known as subitising).

The gingerbread kids (page 24)

It is valuable to link maths to other subjects where possible. This sheet can be used in conjunction with the story of the gingerbread man or with cooking or alternatively art activities, for example potato printing a row of gingerbread men, where buttons are drawn or stuck on after.

SUGGESTED QUESTION:

- How many buttons have you drawn on this gingerbread kid?

Roller coaster ride (page 25)

The teachers' note at the foot of the sheet provides a suggestion for a group starter activity. Some children might find it helpful to count out counters first and then draw that number of children.

SUGGESTED QUESTIONS:

• Guess how many fingers I am holding up.
• Was your guess more or less than the answer?

The three bears (page 26)

In this activity, the children are encouraged to look for objects in a picture and count them. This requires more skill than counting objects in a set because it is easier to count an object more than once or to miss an object altogether.

SOLUTIONS:

There are 3 bowls, 3 spoons, 1 saucepan, 4 photos, 4 chairs and 5 mugs.

SUGGESTED QUESTIONS:

• Did you find this work easy or difficult? Why?
• Are there more bowls or spoons?
• Are there more chairs or mugs?
• Which items are there four of?

Music makers (page 27)

In a movement lesson the children can try the following activity before being given the sheet as a follow-on activity.

As the children are marching randomly around the room pretending to play musical instruments, say the rhyme:

How many members in our band?
How many members in our band?
How many members in our band?
There are * today!

where * is replaced by a number (such as between one and five).

As the number is said, the children should try to get into groups of that size and sit down. The first group to do this are allowed to stand up and do a mini performance by pretending to play their instruments. All join in again and the rhyme is said again.

SUGGESTED QUESTIONS:

• How many members in this band?
• If your whole family were in one band, how many would there be?

In the jungle (page 28)

As for 'The three bears' activity, here the children are encouraged to look for objects in a picture and count them. This requires more skill than counting objects in a set because it is easier to count an object more than once or to miss an object altogether. This activity sheet involves sets of objects up to ten.

SOLUTIONS:

There are 2 jeeps, 5 monkeys, 6 suitcases, 9 parrots, 4 pairs of binoculars and 7 people.

SUGGESTED QUESTIONS:

• Which items are there four of?
• Are there more jeeps or parrots?

Shoe story (page 29)

This activity involves counting, but also begins to explore ideas of addition and subtraction. The sheet can also be enlarged on a photocopier, coloured by the children and used as a classroom display with the numbers masked and questions saying 'How many are red? How many are yellow?' etc. This work with shoes and pairs of shoes can also be linked with ideas of counting in twos.

SUGGESTED QUESTIONS:

• How many more slippers are coloured than boots?

Hunt at home (page 30)

Linking with other topic work about the home, this sheet can be given to the children as an additional activity to do with an adult at home. Alternatively, some of the items could be masked and altered and used for counting objects around the school or in the classroom, for example tables, waste bins.

SUGGESTED QUESTIONS:

• How easy did you find this activity?
• Talk to your partner. Who has more light-switches/tins of food/spoons?
• What do you have two of?

Estimate how many objects they can see and check by counting

Children often find estimating uncomfortable as they may think that inaccuracy in maths is unacceptable. It is important that they are shown that estimating and approximating occurs all the time in everyday life, for example: 'He ate about six sweets' or 'I'll be ready in about 20 minutes'. Watch out for children changing their estimates after they have actually counted, or for those who do not estimate but merely count from the start. When the children are estimating, rather than emphasising how close/far away their estimates are from the actual answer, encourage them to see how checking helps them to become much better at estimating next time.

Countryside creatures (page 31)

Encouraging children to say which sets have 'about five' can help them gain confidence with estimating and with recognising small numbers of objects without the need to count them (known as subitising). At the start of the lesson talk to the children about numbers that are 'about five', and the fact that those sets with four or six etc. can also be described correctly as 'about five', and use sets of practical objects to demonstrate this.

SUGGESTED QUESTIONS:

- How many foxes are there?
- Are there more mice than hares?
- How many more hedgehogs are there than moles?

At the art gallery (page 32)

Introduce this activity by asking the children to estimate and then count a number of cubes or beads. Ensure that the children are familiar with the language of estimating, using vocabulary such as 'guess', 'about', 'roughly' and 'nearly'.

SUGGESTED QUESTIONS:

- How many rabbits are there?
- Are there more horses than rabbits?
- How many more rabbits are there than sheep?

In the kitchen (page 33)

Using a shelf in the classroom with items on, encourage the children to estimate the number, for example 'about ten books', 'about five pots of pencils'. Remember that when children are estimating, rather than emphasising how close/far away their estimates are from the actual answer, you should encourage them to see how checking helps them to become much better at estimating next time.

SUGGESTED QUESTIONS:

- Can you draw a shelf with seven bottles on?
- How many more cups would be needed if ten people came to tea?

The estimating game (page 34)

The sheet could be enlarged on a photocopier, stuck onto card, coloured and laminated for a more permanent classroom resource. As an alternative activity, two sets of the cards could be used to play a 'snap' game. If two cards contain the same number of items the first player to call 'snap' wins the cards.

SUGGESTED QUESTIONS:

- Which sets of items were the easiest to estimate? Why?
- Which card has the most/fewest items?

Count aloud in ones, twos, fives or tens

When children are learning to count on and back in ones, twos, fives or tens it can be helpful to use number lines. Ask a child to point to each number in turn as the class count on and back in ones. Choose another child to point to 'every other number', starting at zero, for counting on and back in twos and so on. Ask the children to say what they notice about the numbers when counting on from zero in fives or tens, drawing attention to the last digit of the numbers.

Count aloud game (page 35)

The children require a cube each and a die. The sheet can be masked and the instructions on the board altered to include counting back in ones, for example, 'Count back from seven to one'. The children can also be asked to count back to zero, rather than one.

SUGGESTED QUESTIONS:

- Have you been listening carefully to see if anyone made a mistake?
- Can you show me how you counted to nine?

Lovely gloves (page 36)

This activity can be used to reinforce counting in ones to 10. It also provides practice in writing or copying numerals. As an alternative to the activity in the Teachers' note, ask the children to place their hands as if playing the piano and tap each finger as a number is said aloud. Some children could be asked to write the numerals on the second pair of gloves from right to left, counting backwards from 10.

SUGGESTED QUESTIONS/PROMPT:

- Close your eyes and count from one to ten, tapping your head as you say each number.
- Which number comes after seven?
- What is the third number you said?
- Can you count backwards from ten to one?

Rise and shine (page 37)

This activity can be used as an assessment tool to see how confident the children are in counting to 15 and recognising and writing the numerals to 15 and beyond.

SUGGESTED QUESTIONS/PROMPT:

- Which number comes after 7?
- Which number comes before 11?
- Now close your eyes and count from 1 to 16, tapping your head as you say each number.

Tea for two (page 38)

Here the children need to be able to write the numbers to 20 in figures. For those children who are less confident, some of the numbers could be written on the mugs before photocopying. By colouring every other mug, the children are being introduced to odd and even numbers, and the extension activity encourages them to write the even numbers. They should be encouraged to say them in order (count in twos).

SUGGESTED QUESTIONS:

- Can you count in twos, starting with 2? How far can you go?
- Which number comes after 8 if you are counting in twos?

Sandwich boxes (page 39)

At the start of the lesson, demonstrate how things can be counted in twos, for example invite some children to stand at the front and explain that you are going to count their ears! Ask the rest of the class to count in twos as you point to each child. Check the answer at the end by counting in ones to show the children that they are correct. Repeat with other sets of children, counting eyes, feet, arms, etc.

SUGGESTED QUESTIONS:

- What patterns did you notice in the numbers?
- How many children will have eight ears between them?
- How many sandwiches are there in five boxes?

Monster eyes (page 40)

As for the previous activity, demonstrate how items can be pointed to as the children count in twos to find the total. Practise counting in twos at the start of the lesson.

SUGGESTED QUESTIONS:

- How far can you count in twos? Can you go any further?
- Which number comes after 12 if you are counting in twos?

Puppies and dogs (page 41)

The children should be encouraged to fill in the missing numbers and then say all the numbers aloud, pointing to each number as they say it. This provides practice of counting in ones and of recognising and writing numbers in figures. Attention should then be drawn to the numbers being spoken by the mummy dogs, to introduce counting in fives. Ask the children to say these numbers aloud.

SUGGESTED QUESTIONS:

- What do you notice about the numbers when you count in fives?
- What is special about the last digit of each number?
- Which number do you think comes after 50, when you count in fives?
- How far can you count in fives?

Dino spikes (page 42)

Before the children complete this activity, check they understand that they have to find the total number of spikes, rather than the number of dinosaurs. Emphasise that counting in fives is a quick way to do this.

SUGGESTED QUESTIONS:

- How could you check this answer? Could you count in ones? Or could you even count in tens?
- Which number comes after 25, when you count in fives?
- How many spikes do nine dinosaurs have?

Kittens and cats (page 43)

The children should be encouraged to fill in the missing numbers and then say all the numbers aloud, pointing to each number as they say it. This provides practice of counting in ones and of recognising and writing numbers in figures.

Attention should then be drawn to the numbers being spoken by the mummy cats, to introduce counting in tens. Ask the children to say these numbers aloud and to look for patterns in the numbers.

SUGGESTED QUESTIONS:

• What do you notice about the numbers when you count in tens?
• What is special about the last digit of each number?
• Which number do you think comes after 50, when you count in tens?
• How far can you count in tens?

10 teeth (page 44)

At the start of the lesson, practise counting in tens and ask individual children to write the numbers on the board or hold up the numbers on cards as you do so.

SUGGESTED QUESTIONS:

• How could you check this answer? Could you count in fives instead?
• Which number comes after 40, when you count in tens?
• How many teeth do eight crocodiles have?

Toe counting (page 45)

The purpose of this activity is to encourage the children to practise counting in tens. For this reason, ten toes are not shown on each pair of feet.

SUGGESTED QUESTIONS:

• If there are five in our group, how many toes are there altogether?
• Which number comes after 70, when you count in tens?

Use language such as 'more' or 'less' to compare two numbers

When comparing two numbers or two uncountable quantities, for example amounts of sand or water, the language 'more' and 'less' or 'larger' and 'smaller' should be used. When comparing two sets of countable objects, however, the words 'more' and 'fewer' should be used.

On the fence (page 46)

This activity involves comparing two sets of birds on fences and saying which fence holds more. Encourage the children to count to find or check their answers. A number line could be used to help them appreciate which number in a pair is larger.

SUGGESTED QUESTIONS:

• How did you know this fence had more? Did you count them?
• How could you check your answers?
• Which fence has fewer birds?

Sea-life world (page 47)

For this activity, ensure the children realise that they must count the number of sea creatures in the picture and compare them. A further extension activity would be to sort all the sea creatures, starting with the creature of which there are fewest: one turtle, two sharks, three jellyfish, four crabs, five seahorses and six fish.

SOLUTIONS:

seahorse	crab
shark	fish
NTT	
jellyfish	turtle
shark	jellyfish

SUGGESTED QUESTIONS:

• Of which creature are there most/fewest?
• How many more seahorses are there than turtles?
• How many fewer crabs are there than fish?

Brothers and sisters (page 48)

Begin the lesson by asking the children to hold up fingers to show how many brothers and sisters they have. Discuss those with no fingers held up and ask a child to write this amount (zero) as a number on the board. Discuss other amounts and ask who has the most brothers and sisters in the class. Ask the children to get into pairs and see who has more brothers and sisters, or whether they have the same number.

SUGGESTED QUESTION:

• Who has more brothers and sisters, Sean or Lucy?

Snail hunt (page 49)

This activity is similar to the previous one in that it involves comparing two numbers without counting objects, but this time using numbers up to ten. You could provide number tracks to ten for children to refer to, if necessary.

SUGGESTED QUESTIONS:

• Which number is more?
• How many more is it?
• Which pot shows the largest number?

Football scores (page 50)

The children could count out counters to represent the goals and compare these. If appropriate, discussion on scores that include the word 'nil' could take place, showing where the zero is in relation to the other numbers and showing the symbol for zero.

SUGGESTED QUESTION:

• This team scored more goals. Which team scored fewer goals?

Baby, baby: 1 and 2 (pages 51–52)

These worksheets could be enlarged on a photocopier, stuck onto card, coloured and laminated for a more permanent classroom resource. Children find this game very entertaining once they are comfortable with the rules. It allows them to compare numbers (comparing two numbers if two play, three numbers if three play, and so on).

Additional rule If, for example, red is chosen and two players each have three red jelly babies and no other players have a higher number of red jelly babies, then green or yellow should be chosen and the higher of the two wins the cards.

SUGGESTED QUESTIONS:

• What did you like best about the game?
• Can we make some more cards like this to use?

Compare jars (page 53)

To emphasise the number in the jars, the children could count counters into pots and label them. They could then compare the number of counters by matching.

SUGGESTED QUESTION:

• How do you know this jar has more?

More or less (page 54)

This activity involves a more abstract use of number, where the numbers on the cards do not represent a number of objects. Encourage the children to use a number line for this activity. When comparing the numbers, the children should use 'less' for the smaller number, rather than 'fewer'. Children who finish the extension activity quickly, could write or say as many numbers as they know that are more than 3.

SUGGESTED QUESTION/PROMPT:

• Tell me a number that is more/less than 5.
• How do you know that 10 is more than 4?

Ship race (page 55)

In this activity, the children compare numbers, rather than sets of countable objects or events. When doing so, they are asked to use 'larger' and 'smaller'. Children who find this difficult could count out numbers of objects and then compare the sets.

SUGGESTED QUESTIONS:

• Would you rather have 5 sweets or 8 sweets?
• How do you know that 3 is smaller than 7?

Use ordinal numbers in different contexts

Ordinal numbers are those that we use to describe the order of a set of items: first, second, third... They are closely linked to the counting numbers 'one', 'two, 'three'... At this stage the children need to learn the words and relate them to the numbers. Through experience, the children will begin to understand and learn the notation 'st', 'nd', 'rd', 'th', and know when to use each one, for example writing '5th', not '5rd'.

Biggest is best (page 56)

This first ordinal number activity involves only 1st, 2nd and 3rd in the context of prizes. The children are required to position the vegetables correctly on the chart. This can be practised at the start of the lesson using sets of objects such as plants, pencils, etc.

SUGGESTED QUESTIONS:

- Which is the biggest carrot?
- The smallest carrot will get which prize?

Go-kart race (page 57)

On the board, write this list in order: 1st, 2nd, 3rd, 4th, 5th, 6th, 7th and 8th. Talk about what these mean and ask the children to say them aloud. Invite children to stand in a line and call out who is first, second, third, etc. Show a selection of items in a line and ask the children to identify which is third, fifth, second, etc. The children will require coloured pencils for this activity.

SUGGESTED QUESTIONS:

- What colour is the third go-kart?
- What colour is the fifth go-kart?
- What colour is the fourth go-kart?

Dumper trucks (page 58)

This activity explores ordinal numbers up to tenth position. As with the previous activity, the children will need a range of coloured pencils.

SUGGESTED QUESTIONS/PROMPT:

- What colour is the seventh truck?
- What colour is the fifth truck?
- Colour the ninth truck in a different colour.

Lining-up time (page 59)

The children will need a long strip of paper on which to stick the strips and cards. To provide a colourful display, the children could colour the people and the cards and stick them onto coloured paper.

SUGGESTED QUESTIONS:

- Which child comes after the child in fifth position?
- Is the child in third place wearing spectacles?
- In which position will you find a child wearing a scarf or a hat?

Recognise numerals 1 to 9

There are many ways in which children can be helped to recognise and begin to write numerals. Ask the children to trace numbers in the air, in sand, on paper, on textured material, on the palm of their other hand, on a friend's back, etc. Different types of fabric (such as sandpaper, corrugated paper, textured wallpaper) can help children to gain a 'feel' for the shape of the numeral. Describing the shapes of the numerals in words is also very useful, for example 'straight across and then down' (for 7), or 'start at the top and draw an egg shape' (for 0). The children can also be asked to make numerals out of Plasticine, or paint them on paper to display.

Cheeky chimps (page 60)

Ensure that the numbers 0 to 9 are displayed in the classroom. A similar activity can be performed practically with large plastic numerals, using a piece of paper or a hand to mask parts of the numeral. Encourage the children to explain their reasoning throughout this type of activity: for example, 'These parts are curved, so it must be one of the curved numbers, like 3 or 8.'

SUGGESTED QUESTIONS:

- Which number do you think it is?
- What makes you think it must be a 3?
- Do any numbers have only straight sides?
- Which numbers are curved?

Find the numbers (page 61)

The focus of this activity is on recognising numerals amongst other shapes and letters. The extension activity can lead to a visually stimulating resource.

SUGGESTED QUESTIONS:

- Is this a number or a letter?
- Are you sure you have found all of the numbers?

Tracing tracks, Tracing tractors, Tracing trainers (pages 62–64)

These three worksheets can be used to help those children who are experiencing difficulty with writing numerals correctly. Ensure that the children understand how to write the numerals by starting at the dot. Show how for numerals 4 and 5 they must lift their pencil off the paper and move it for the next part of the numeral. Make sure that the children are not just blindly copying without thinking about which number it is and what it stands for.

SUGGESTED QUESTIONS:

- Which number are you writing?
- Show me that many fingers.
- How many of the number 5 are on this page?

Using the CD-ROM

The PC CD-ROM included with this book contains an easy-to-use software program that allows you to print out pages from the book, to view them (e.g. on an interactive whiteboard) or to customise the activities to suit the needs of your pupils.

Getting started

It's easy to run the software. Simply insert the CD-ROM into your CD drive and the disk should autorun and launch the interface in your web browser.

If the disk does not autorun, open 'My Computer' and select the CD drive, then open the file 'start.html'.

Please note: this CD-ROM is designed for use on a PC. It will also run on most Apple Macintosh computers in Safari however, due to the differences between Mac and PC fonts, you may experience some unavoidable variations in the typography and page layouts of the activity sheets.

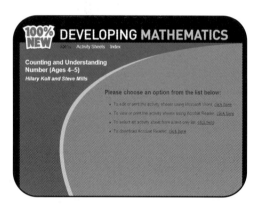

The Menu screen

Four options are available to you from the main menu screen.

The first option takes you to the Activity Sheets screen, where you can choose an activity sheet to edit or print out using Microsoft Word.

(If you do not have the Microsoft Office suite, you might like to consider using OpenOffice instead. This is a multi-platform and multi-lingual office suite, and an 'open-source' project. It is compatible with all other major office suites, and the product is free to download, use and distribute. The homepage for OpenOffice on the Internet is: www.openoffice.org.)

The second option on the main menu screen opens a PDF file of the entire book using Adobe Reader (see below). This format is ideal for printing out copies of the activity sheets or for displaying them, for example on an interactive whiteboard.

The third option allows you to choose a page to edit from a text-only list of the activity sheets, as an alternative to the graphical interface on the Activity Sheets screen.

Adobe Reader is free to download and to use. If it is not already installed on your computer, the fourth link takes you to the download page on the Adobe website.

You can also navigate directly to any of the three screens at any time by using the tabs at the top.

The Activity Sheets screen

This screen shows thumbnails of all the activity sheets in the book. Rolling the mouse over a thumbnail highlights the page number and also brings up a preview image of the page.

Click on the thumbnail to open a version of the page in Microsoft Word (or an equivalent software program, see above.) The full range of editing tools are available to you here to customise the page to suit the needs of your particular pupils. You can print out copies of the page or save a copy of your edited version onto your computer.

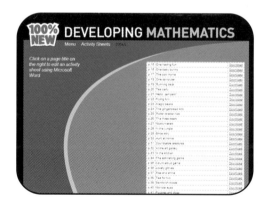

The Index screen

This is a text-only version of the Activity Sheets screen described above. Choose an activity sheet and click on the 'download' link to open a version of the page in Microsoft Word to edit or print out.

Technical support

If you have any questions regarding the *100% New Developing Literacy* or *Developing Mathematics* software, please email us at the address below. We will get back to you as quickly as possible.

educationalsales@acblack.com

One having fun

• **Match the number to the children.**

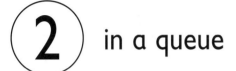 1 having fun

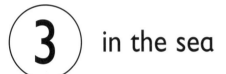 2 in a queue

3 in the sea

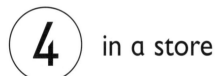 4 in a store

 5 by a hive

 **NOW TRY THIS!**

• **Write a number that is** less than 3 .

• **Write a number that is** more than 5 .

Teachers' note Say the rhyme several times at the start of the lesson until the children are familiar with the words. Encourage them to think up other lines for the rhyme, beginning with 6, 7, 8 and so on, such as '6 near some bricks'.

100% New Developing Mathematics
Counting and Understanding
Number: Ages 4–5
© A & C BLACK

One baby bunny

One baby bunny playing in the sun

One baby bunny playing in the sun

And if one more baby bunny comes to have some fun

There'll be two baby bunnies playing in the sun…

• How many bunnies on the grass?

 1

 **NOW TRY THIS!**

• Draw bunnies on the grass to match each number.

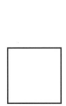 3

 2

1

4

Teachers' note Begin the lesson by repeating the rhyme to the same tune as the song 'Ten green bottles', increasing the number of bunnies by one each time.

100% New Developing Mathematics
Counting and Understanding
Number: Ages 4–5
© A & C BLACK

The zoo rhyme

(1) (5) (7) (9) (8) (2) (3) (6) (4) (10)

• **Fill in the missing numbers.**

(1) ( )   let's go to the zoo!

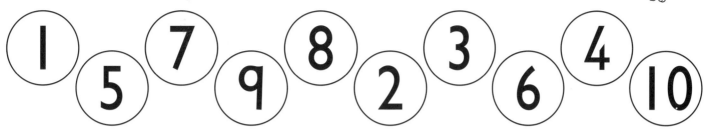

( ) ( )   hear the tigers roar!

( ) ( )   watch the monkeys' tricks!

( ) ( )   feeding time is late!

( ) ( )   see the lions in their den!

NOW TRY THIS!

• **Colour the number** 2 **in the rhyme.**
• **Miss a number. Colour the next number.**
• **Continue to the end.**

Teachers' note Encourage the children to use actions as they say the rhyme. You could repeat the rhyme ten times, each time putting your finger on your lips instead of saying a number name, beginning first by missing 'one', then 'one' and 'two', and so on. Encourage the children to 'think' the missing number names in their heads.

100% New Developing Mathematics
Counting and Understanding
Number: Ages 4–5
© A & C BLACK

One computer...

One computer, two computers,

three computers, four.

Five computers, six computers,

seven computers, more!

• How many computers on each card?

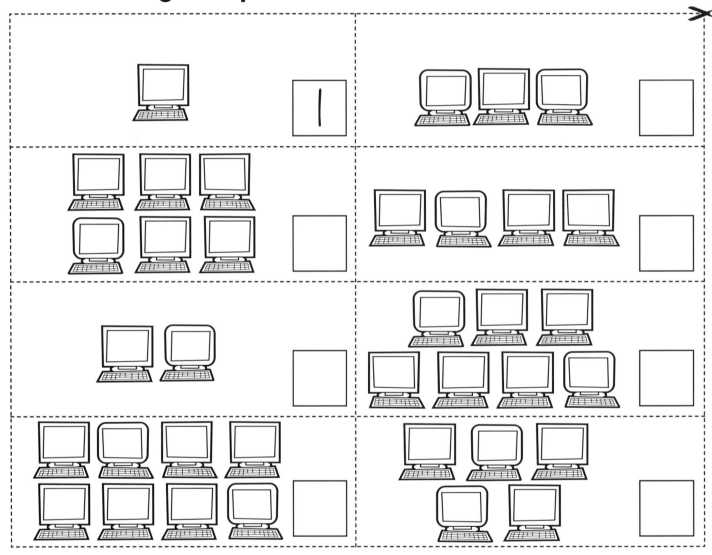

NOW TRY THIS!

- • Cut out the cards.
- • Put them in order from smallest to largest.

Teachers' note This rhyme is an adaptation of 'One potato, two potatoes'. If the children have access to more computers, the rhyme can be used to count some of them. In the extension activity, some children may need help cutting out the cards.

**100% New Developing Mathematics
Counting and Understanding
Number: Ages 4–5
© A & C BLACK**

18

• Play this game with a friend.

☆ Take turns to spin the spinner and move your cube forward.

☆ Say the numbers from 1 to the number you land on.

☆ If you are right collect a medal.

☆ The winner is the first player to collect 6 medals.

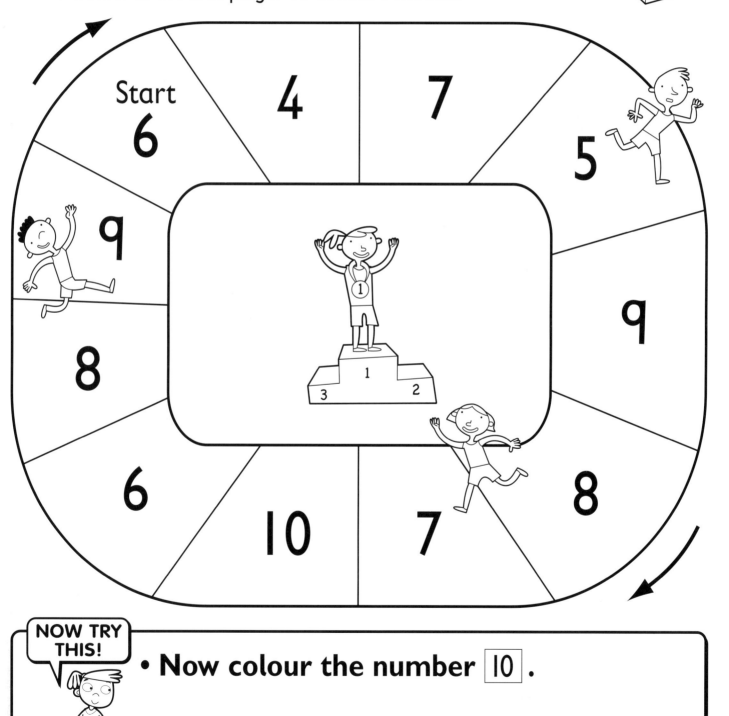

Start
6

4

7

5

9

9

8

8

6

10

7

NOW TRY THIS!

• Now colour the number ⬚10 .

Teachers' note Children can play this game in pairs or small groups with an adult. Explain that cubes for each child are placed on 'Start' and that they spin the spinner and move on from this position for their first go. The numbers on the track can be masked and made larger to provide a more advanced game, for example with numbers to 20. Foil wrap some counters for the medals.

100% New Developing Mathematics
Counting and Understanding
Number: Ages 4–5
© A & C BLACK

Tea party

• **How many teabags in each teapot?**

• **Colour the teapot with** $\boxed{4}$ **bags.**

Teachers' note This activity could be used as a follow-on activity to performing the action rhyme 'I'm a little teapot, short and stout'. During play activities, the children can pretend to make cups of tea in a pot counting out teabags into a pot. The vocabulary 'one more' can be used as part of these activities.

**100% New Developing Mathematics
Counting and Understanding
Number: Ages 4–5
© A & C BLACK**

20

• **How many tents?**

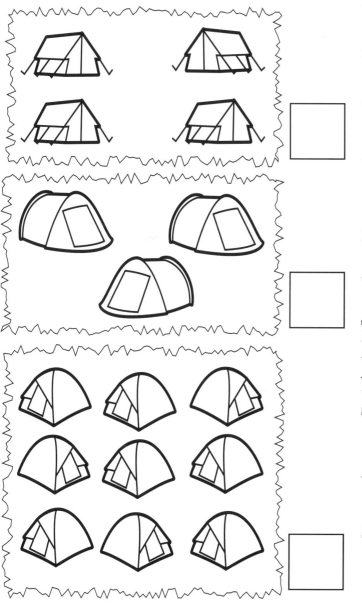

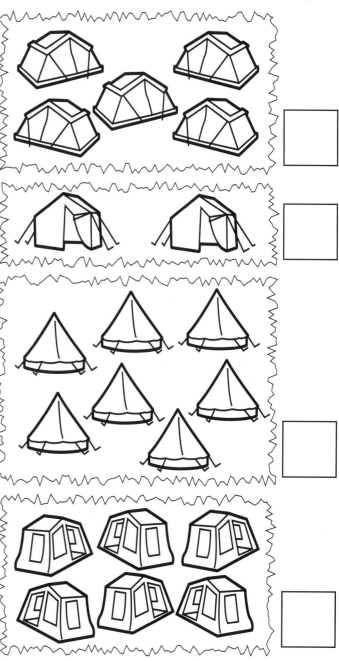

NOW TRY THIS!

• **Draw the same number of tents.**

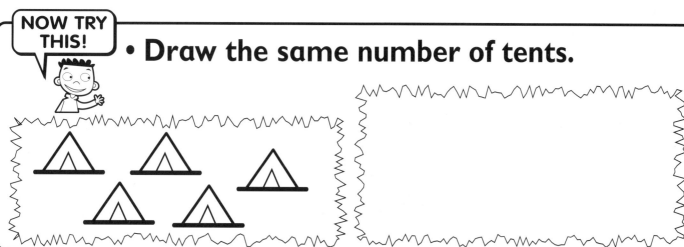

Teachers' note This activity can be used as an assessment tool to see how carefully children can count sets of objects. When counting, some children may find it easier to cover each tent in a set with a cube or counter and then count those.

100% New Developing Mathematics Counting and Understanding Number: Ages 4–5
© A & C BLACK

Frying fun!

• **How many eggs in each pan?**

• **Colour the pan with** 8 **eggs.**

Teachers' note This activity can be used as an assessment tool to see how carefully children can count sets of objects. Some children may find it easier to cover each egg in a set with a cube or counter and then count those.

**100% New Developing Mathematics
Counting and Understanding
Number: Ages 4–5**
© A & C BLACK

Magic beans

• **How many beans in each hand?**

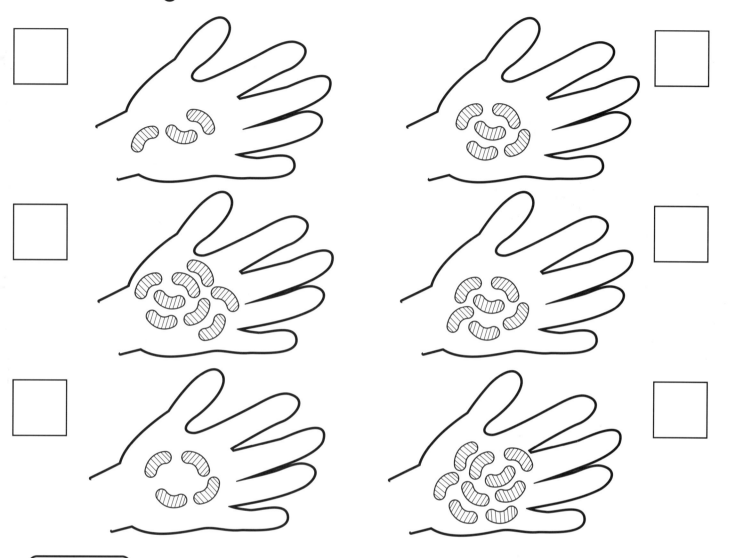

NOW TRY THIS!

• **Draw the same number of beans in the other hand.**

• **How many beans are there** [altogether] **?**

Teachers' note Encourage the children to say the numbers aloud as they count the beans. They could put a tick or a cross on each bean as they count it. Once completed, they could be asked to swap sheets to check their friend's answers.

**100% New Developing Mathematics
Counting and Understanding
Number: Ages 4–5
© A & C BLACK**

23

The gingerbread kids

- Join each gingerbread kid with a number to show how many buttons it has.

- Colour the kid with ⎡4⎤ buttons ⎡blue⎤.
- Colour the kid with ⎡2⎤ buttons ⎡red⎤.
- Colour the kid with ⎡3⎤ buttons ⎡yellow⎤.

NOW TRY THIS!

- Draw buttons to match the numbers.

Teachers' note This activity can be linked with the story of the gingerbread man or with cooking activities or imaginary play. The activity encourages the children to begin to recognise small numbers of items without the need to touch each item.

100% New Developing Mathematics
Counting and Understanding
Number: Ages 4–5
© A & C BLACK

Roller coaster ride

- **Write the number of children in each car.**

- **Colour the cars with** ⟦3⟧ **children** ⟦blue⟧.
- **Colour the cars with** ⟦2⟧ **children** ⟦red⟧.

NOW TRY THIS!

- **Draw the number of children to match each number.**

Teachers' note At the start of the lesson, play a guessing game counting up to 5 small objects. Hold up a piece of card to represent a car on a roller coaster. With your hand behind the card hold up some fingers (to represent people) and ask a child to guess how many fingers you are holding up. After the guess move your fingers up to reveal them and ask the children whether the guess was correct or not.

100% New Developing Mathematics
Counting and Understanding
Number: Ages 4–5
© A & C BLACK

25

The three bears

• **Write how many**

bowls ☐

saucepans ☐

chairs ☐

spoons ☐

photos ☐

mugs ☐

NOW TRY THIS!

• **Colour** ☐ 1 ☐ **mug** green .
• **Colour** ☐ 4 ☐ **mugs** red .
• **Colour** ☐ 2 ☐ **bowls** yellow .

Teachers' note This activity can be used as a follow-on to telling the story of *Goldilocks and the three bears*. Some children may find it easier to cover items of a particular type, for example chairs, with cubes or counters and then count them as they are removed. Alternatively, those who frequently miscount could be encouraged to cross off items as they count.

100% New Developing Mathematics
Counting and Understanding
Number: Ages 4–5
© A & C BLACK

Music makers

• **Write the number of people in each band.**

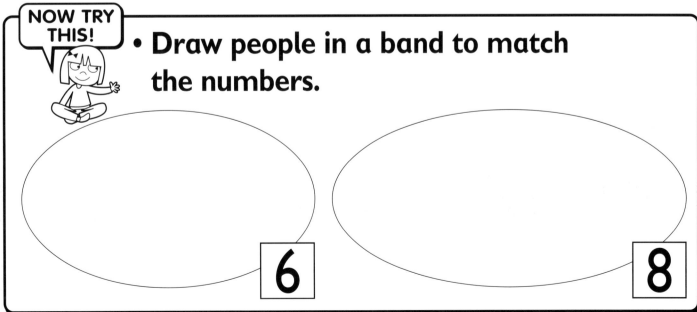

NOW TRY THIS!

• **Draw people in a band to match the numbers.**

6

8

Teachers' note This activity can be used after a music or movement lesson to extend the theme of using and playing musical instruments in a band. It can be used as an assessment activity to see how carefully children can count sets of objects and write numerals to match the number.

**100% New Developing Mathematics
Counting and Understanding
Number: Ages 4–5**
© A & C BLACK

In the jungle

• Write how many

jeeps  □

suitcases □

binoculars □

monkeys  □

parrots □

people  □

NOW TRY THIS!

- **Colour** `2` **jeeps** `green`.
- **Colour** `4` **parrots** `blue`.
- **Colour** `3` **parrots** `yellow`.

Teachers' note This activity can be used to assess how carefully the children can count sets of objects and write the related numerals. Some children may find it easier to cover items with cubes or counters and then count them as they are removed. Those who frequently miscount could be encouraged to cross off items as they count. Children require coloured pencils for the extension activity.

100% New Developing Mathematics
Counting and Understanding
Number: Ages 4–5
© A & C BLACK

Shoe story

- **Colour** 4 **trainers** blue .
- **How many are not coloured?** ☐

- **Colour** 5 **boots** red .
- **How many are not coloured?** ☐

- **Colour** 6 **skates** yellow .
- **How many are not coloured?** ☐

- **Colour** 7 **slippers** pink .
- **How many are not coloured?** ☐

NOW TRY THIS!
- **How many altogether?**

  ☐ ☐  ☐ 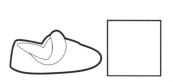 ☐

Teachers' note This activity can be used as an assessment tool to see how carefully children can colour the correct number of objects in a set for numbers up to 10. Children will require coloured pencils for this activity. It may be useful for children to tick or circle the correct number of shoes first and then colour them.

100% New Developing Mathematics
Counting and Understanding
Number: Ages 4–5
© A & C BLACK

29

Hunt at home

- **How many in your home?**
- **Write the number on the line.**

taps	doors	TVs	teddies

  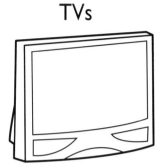

___   ___   ___   ___

sofas	beds	lamps	steps

  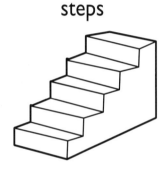 

___   ___   ___   ___

tins of food	spoons	armchairs	light-switches

___   ___   ___   ___

NOW TRY THIS!

- **Find something that you have** 2 **of at home.**
- **Draw them.**

Teachers' note This activity can be given as a home activity for the children to complete with an adult. Encourage the adult to help the child count and write the number of each item. There is no need to count past ten at this stage. If there are none of an item at home, the symbol 0 can be introduced and described as 'zero' or 'none'.

100% New Developing Mathematics
Counting and Understanding
Number: Ages 4–5
© A & C BLACK

Countryside creatures

- **Tick** ✓ **all the sets that you think show** about 5 **creatures.**
- **Then count and write how many.**

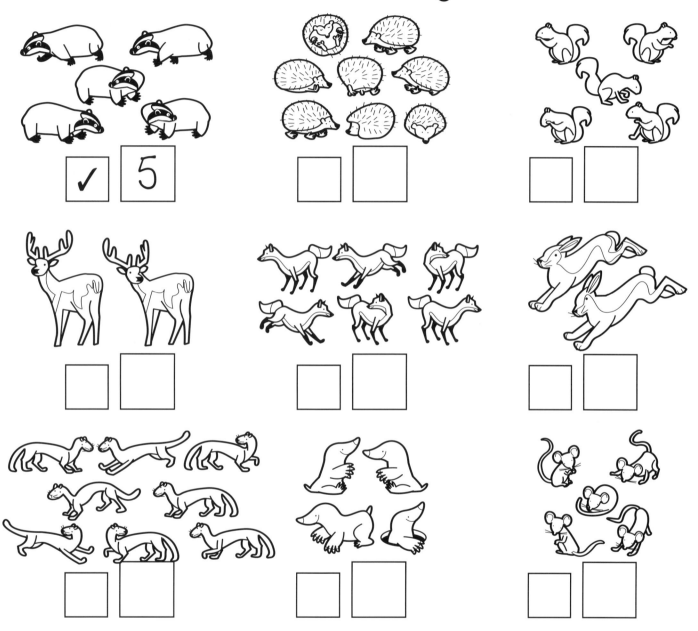

✓ | 5

Teachers' note Remind the children that estimating is not the same as counting. Ask them to use a coloured pencil to quickly tick any sets that have about 5 creatures in and then use a normal pencil to count the sets. Emphasise the importance of looking and making a rough guess, and remind them that it is not just sets that have exactly 5 but those with 4 or 6 that can be described correctly as 'about 5'.

100% New Developing Mathematics Counting and Understanding Number: Ages 4–5
© A & C BLACK

31

At the art gallery

• **Estimate how many animals are in each picture.**

• **Now count them.**

 **NOW TRY THIS!**

• **Estimate how many birds are in each picture.**

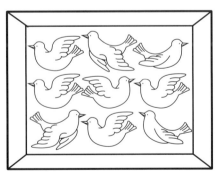  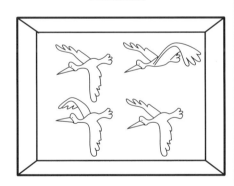

• **Now count them.**

Teachers' note Remind the children that estimating is not the same as counting. Emphasise the importance of looking and making a rough guess, perhaps asking themselves whether it looks like there might be more than 5 or less than 5 to help them estimate.

100% New Developing Mathematics
Counting and Understanding
Number: Ages 4–5
© A & C BLACK

In the kitchen

• **Estimate how many of each item are on the shelves.**

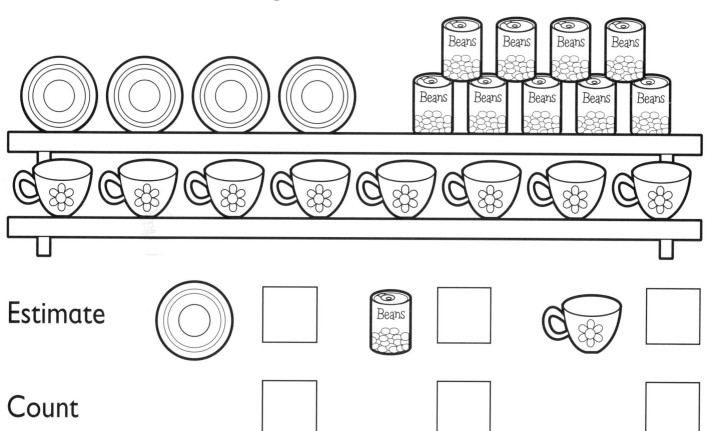

Estimate

⊙ □ 🥫 □ ☕ □

Count

□ □ □

NOW TRY
THIS!

• **How many bottles are on this shelf?**

Estimate □

Count □

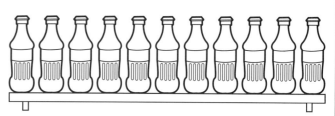

• **Colour** 5 **bottles** green .
• **Colour** 2 **bottles** red .
• **Colour** 3 **bottles** yellow .

• **How many have not been coloured?** □

Teachers' note Ensure the children understand that when estimating they should not actually count the items. Watch out for children who try to change their estimates and emphasise that it is not vital for the estimate to be exactly the same as the number of items.

100% New Developing Mathematics
Counting and Understanding
Number: Ages 4–5
© A & C BLACK

The estimating game

• **Play this game with a friend.**

☆ Cut out the cards. Spread them out face down.

☆ Take turns to pick a card.

☆ Estimate the number of items on the card. Count to check.

☆ If you were right, keep the card. If not, put it back.

☆ The winner is the one with the most cards.

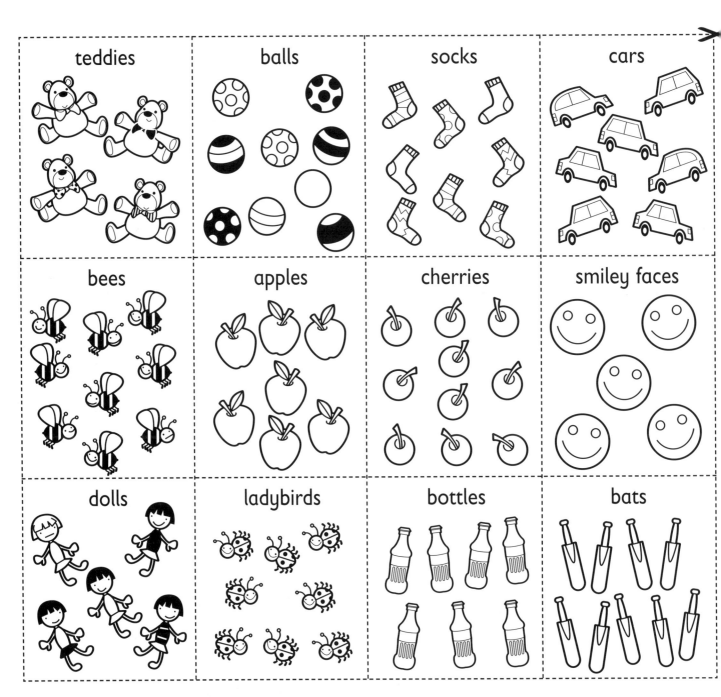

teddies	balls	socks	cars
bees	apples	cherries	smiley faces
dolls	ladybirds	bottles	bats

Teachers' note Discuss the different sizes of things and encourage the children to realise that 'bigger things' does not necessarily mean a larger number of things. As an extension activity, the children could make their own version of the game with up to 20 items on each card or could pick pairs of cards and say which is more, less or whether they are the same.

**100% New Developing Mathematics
Counting and Understanding
Number: Ages 4–5
© A & C BLACK**

Count aloud game

- ## Play this game with a friend.

☆ Each player puts a cube on 'Start'.

☆ Take turns to roll the dice and move your cube forward.

☆ Count aloud from 1 to the number shown.
  If you get it right, collect a counter.

☆ The winner is the first player with 10 counters.

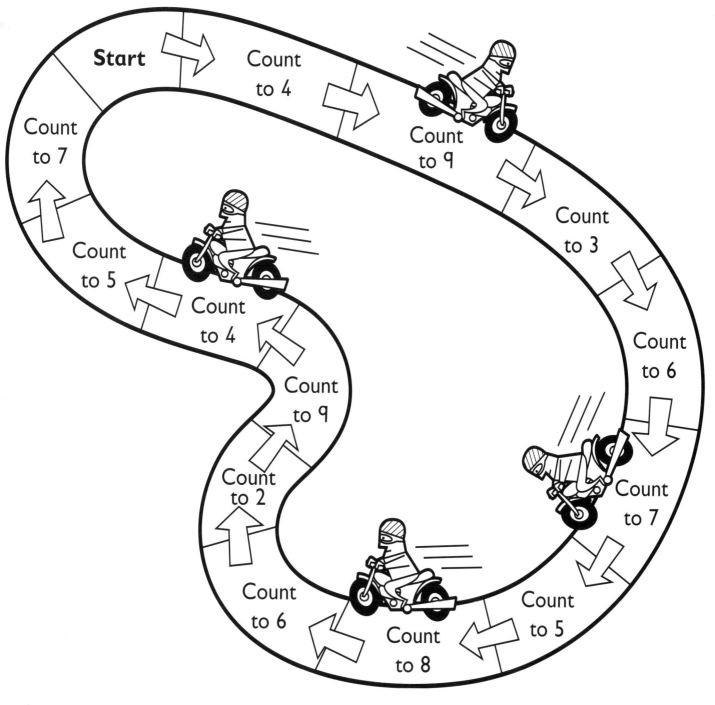

Start

Count to 4

Count to 9

Count to 7

Count to 3

Count to 5

Count to 4

Count to 6

Count to 9

Count to 2

Count to 7

Count to 6

Count to 8

Count to 5

Teachers' note Provide counters for the children to 'win'. This game can also be played in a small group with an adult or can be given as a homework game. It provides an opportunity for the children to listen to each other counting aloud and to spot any mistakes. If an adult is playing, it is sometimes interesting to intentionally make a mistake and see whether the child notices.

**100% New Developing Mathematics
Counting and Understanding
Number: Ages 4–5**
© A & C BLACK

35

Lovely gloves

1 2 3 4 5 6 7 8 9 10

- Count aloud to 10.
- Write $\boxed{1}$ to $\boxed{10}$ in order on each pair of gloves.

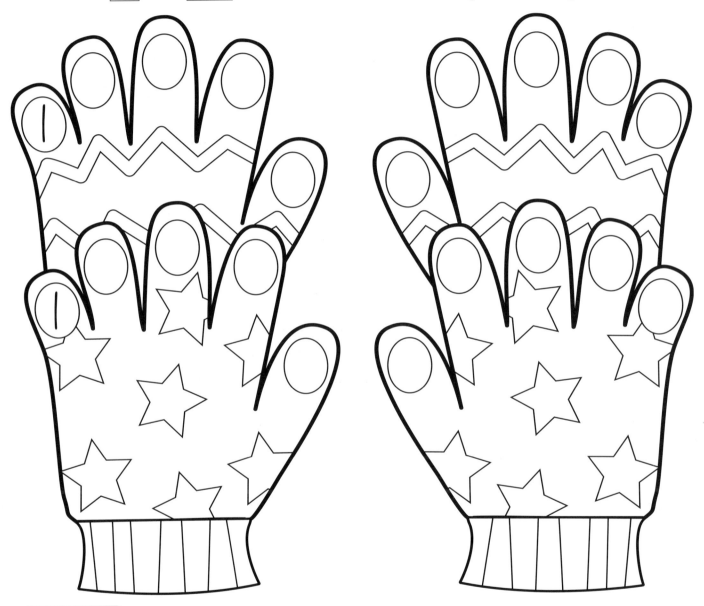

NOW TRY THIS!

- **How many fingers are there altogether?**

10 and 10 is $\boxed{}$

Teachers' note At the start of the lesson, revise counting the numbers from 1 to 10 in different ways. Try whispering the numbers, then shouting them or speaking in a squeaky voice etc. For the kinaesthetic learners, invite the children to hold their hands up, palms outwards, and to touch their nose to each finger as they count from 1 to 10, starting at the left. Children will find this quite difficult!

100% New Developing Mathematics
Counting and Understanding
Number: Ages 4–5
© A & C BLACK

Rise and shine

- **Count aloud from $\boxed{1}$ to $\boxed{16}$. Join the dots in order.**
- **Start with $\boxed{1}$.**

NOW TRY THIS!

- **Count on from 10. Write the numbers.**

12

11

10

- **Now join the dots in order.**

Teachers' note Make sure the children say the numbers aloud as they join the dots. If they are not confident with numerals to 10 and beyond, provide a number line or track. The children could cover each number on the line or track as they count along.

100% New Developing Mathematics
Counting and Understanding
Number: Ages 4–5
© A & C BLACK

37

Tea for two

- **Write the numbers in order on the mugs.**

- **Colour mug 2** `red`**. Miss a mug.**
- **Colour the next mug red. Continue to the end.**

NOW TRY THIS!

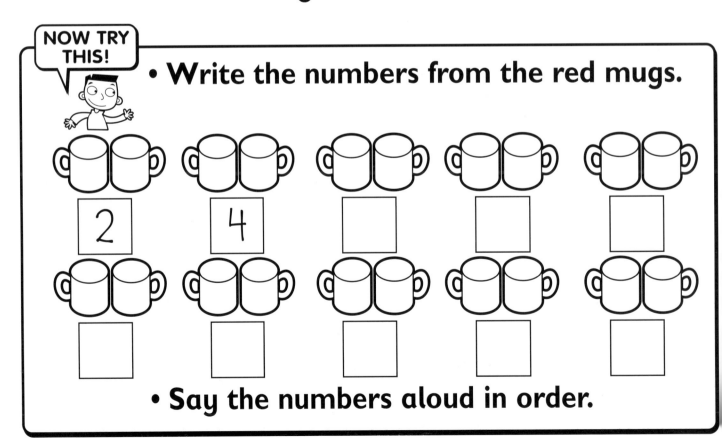

- **Write the numbers from the red mugs.**

| 2 | 4 | | | |

- **Say the numbers aloud in order.**

Teachers' note This activity encourages pupils to realise that, when counting on in ones, every other number is the same as counting in twos. Begin the lesson by counting in ones, whispering one word and shouting the next: one, TWO! three, FOUR!... and then just shouting every other word: TWO! FOUR! SIX!... Explain that this is known as counting in twos.

38

**100% New Developing Mathematics
Counting and Understanding
Number: Ages 4–5**
© A & C BLACK

Sandwich boxes

There are ⬚**2** sandwiches in each box.
- Count in ⬚**2s** to find how many sandwiches in each row.

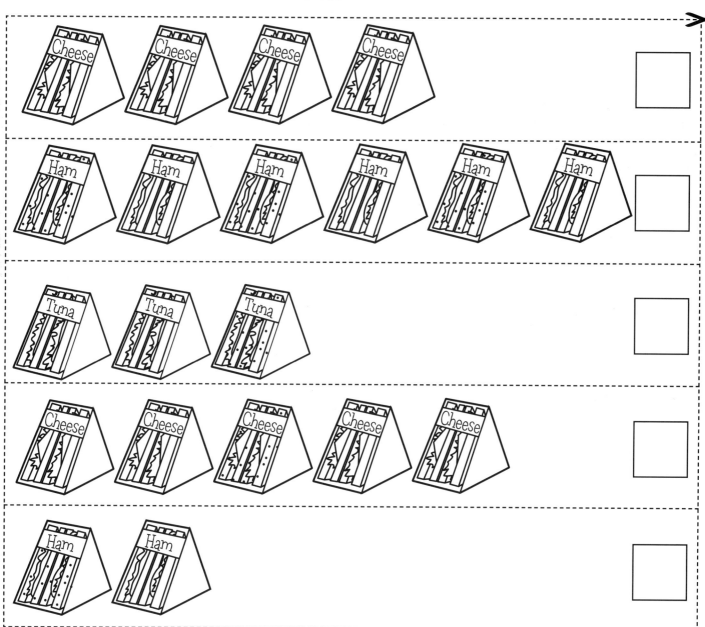

NOW TRY THIS!

- **Cut out the cards.**
- **Put them in order, starting with the ⬚fewest .**

Teachers' note Practise counting in twos at the start of the lesson. Encourage the children to point to each sandwich pack as they count in twos to find the total number of sandwiches in each row. Some children might find it helpful to place two cubes or counters on each sandwich pack and count them to check their answers. In the extension activity, scissors will be needed to cut out the cards.

100% New Developing Mathematics
Counting and Understanding
Number: Ages 4–5
© A & C BLACK

39

Monster eyes

Each monster has 2 eyes.
- Count in 2s to find the number of eyes in each row.

NOW TRY THIS!

- On the back of this sheet, draw 8 monsters each with 2 eyes.
- Write how many eyes altogether.

Teachers' note At the start of the lesson, revise counting every other number by using the 'whisper, SHOUT' approach, for example: one, TWO, three, FOUR, and so on. Then continue this but change whisper to silence, where the number is thought in the head rather than said aloud, to produce only the even numbers TWO, FOUR, SIX…

40

100% New Developing Mathematics
Counting and Understanding
Number: Ages 4–5
© A & C BLACK

- **Fill in the missing numbers.**
- **Colour the numbers said by the mummy dogs.**

NOW TRY THIS!

- **Count in** 5s **. Write the next 5 numbers.**
- **Say the numbers.**

5	10	15	20	25					

Teachers' note At the start of the lesson, ask the children to place one hand on the table (or floor). Ask them to count in ones and to tap fingers as they do so, always starting from the little finger. Each time they tap their thumb on the table they should shout the number. Once children begin to recognise every fifth number, these numbers can be said as a string: 'Five, ten, fifteen…'.

100% New Developing Mathematics
Counting and Understanding
Number: Ages 4–5
© A & C BLACK

Dino spikes

These dinosaurs have ⬚5⬚ spikes each.

- Count in ⬚5s⬚ as you point to each dinosaur.
- Write the number of spikes in each family.

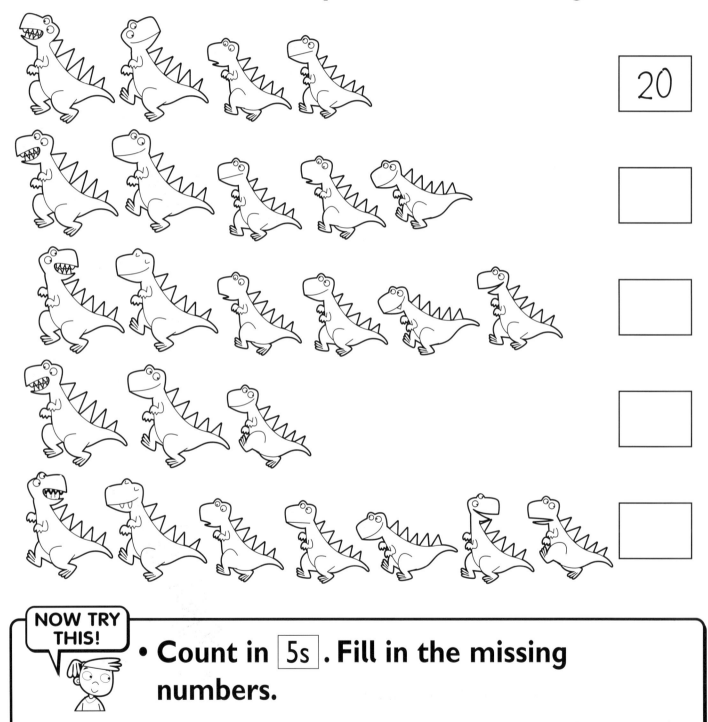

| 20 |

NOW TRY THIS!

- Count in ⬚5s⬚. Fill in the missing numbers.

| 5 | 10 | | | 25 | | | 40 | | |

Teachers' note Practise counting in fives and encourage the children to notice the pattern of the numbers ending in 5 or 0. Invite several children to stand in a row with their palms facing forwards. Show how, by pointing to each hand as you count in fives, you can quickly find the number of fingers being held up. Count the fingers individually to prove this.

100% New Developing Mathematics
Counting and Understanding
Number: Ages 4–5
© A & C BLACK

Kittens and cats

- **Fill in the missing numbers.**
- **Colour the numbers said by the mummy cats.**

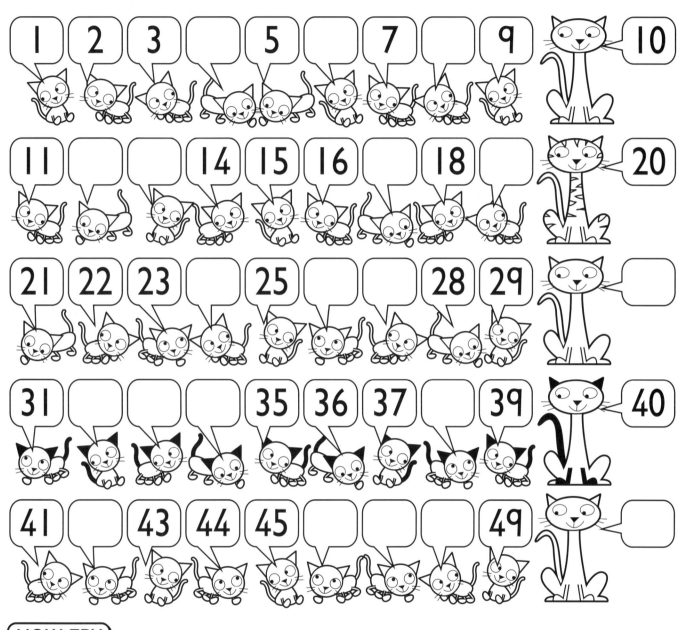

| 1 | 2 | 3 | | 5 | | 7 | | 9 | | 10 |

| 11 | | | 14 | 15 | 16 | | 18 | | | 20 |

| 21 | 22 | 23 | | 25 | | | 28 | 29 | | |

| 31 | | | 35 | 36 | 37 | | 39 | | 40 |

| 41 | | 43 | 44 | 45 | | | 49 | | |

NOW TRY THIS!

- **Count in** [10s] **. Write the next 5 numbers.**
- **Say the numbers.**

| 10 | 20 | 30 | 40 | 50 | | | | |

Teachers' note At the start of the lesson, practise counting in tens, drawing attention to the fact that each number ends in a zero when written in figures. Invite several children to stand in a row with their palms facing forwards. Show how, by pointing to each pair of hands as you count in tens, you can quickly find the number of fingers being held up. Count the fingers individually to prove this.

100% New Developing Mathematics
Counting and Understanding
Number: Ages 4–5
© A & C BLACK

10 teeth

These crocodiles have $\boxed{10}$ teeth each.

- Count in $\boxed{10s}$ as you point to each crocodile.
- Write the number of teeth in each family.

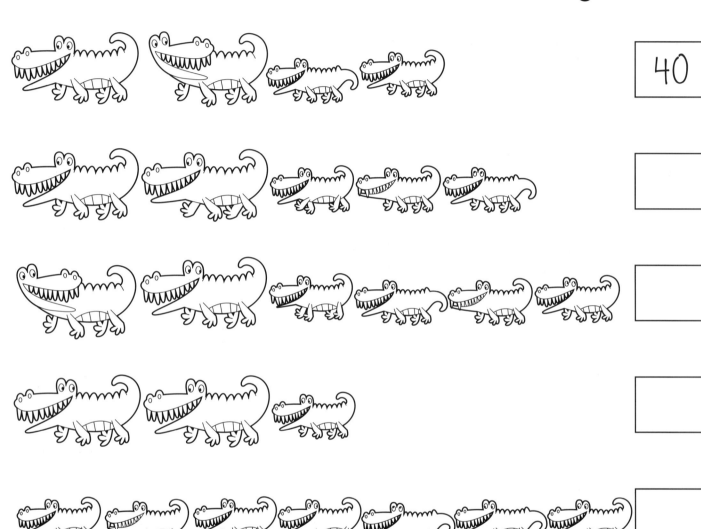

	40

NOW TRY THIS!

- Count in $\boxed{10s}$. Fill in the missing numbers.

10	20			50			80		

Teachers' note Practise counting in tens and encourage the children to notice the pattern of the numbers ending in 0. Listen for children who use the word 'ten-ty', rather than 'a hundred'. Invite several children to stand in a row with their palms facing forwards. Show how, by pointing to each person as you count in tens, you can quickly find the number of fingers being held up.

100% New Developing Mathematics
Counting and Understanding
Number: Ages 4–5
© A & C BLACK

Toe counting

- **Point to each pair of feet and count in** 10s .
- **Write how many toes are in each row.**

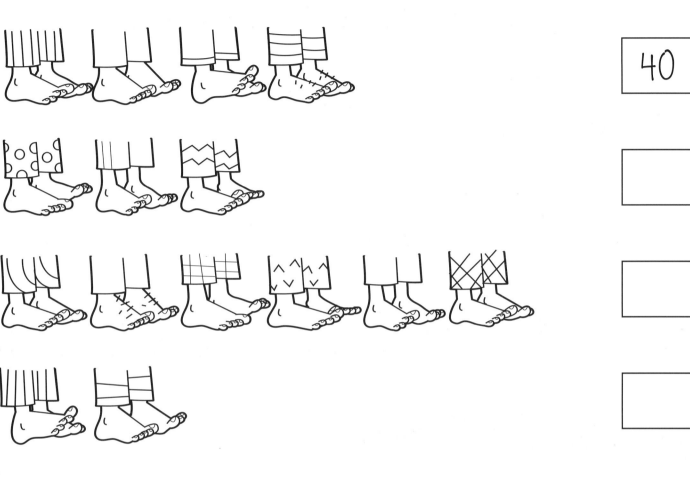

40

NOW TRY THIS!

- **Count in** 10s . **Fill in the missing numbers.**

10   50   100

Teachers' note Practise counting in tens and encourage the children to notice the pattern of the numbers ending in 0. Listen for children who use the word 'ten-ty', rather than 'a hundred'. Invite several children to stand in a row. Remind them that each pair of feet has ten toes and show how, by pointing to each person as you count in tens, you can quickly find the number of toes altogether.

**100% New Developing Mathematics
Counting and Understanding
Number: Ages 4–5
© A & C BLACK**

On the fence

On each fence there are some birds.

• For each pair, colour the fence that has more .

NOW TRY THIS!

• Draw more than 4 birds on this fence.

Teachers' note This activity encourages the children to compare numbers of items when they can be counted. Where children are unable to see which fence has more birds, invite them to count the birds on the fence and compare the numbers, using a number line.

100% New Developing Mathematics
Counting and Understanding
Number: Ages 4–5
© A & C BLACK

Sea-life world

- **Draw a ring around the correct picture.**
- **Are there** | more |

 or ?

 or  ?   or  ?

 or  ?   or ?

NOW TRY THIS!

- **Are there** | fewer |

 or ?    or  ?

 or ?    or  ?

Teachers' note At the start of the lesson, use the words 'more' and 'fewer' in a range of contexts. Note that 'fewer' is used when referring to countable objects, whereas 'less' is used when referring to numbers or uncountable quantities, for example water or sand.

**100% New Developing Mathematics
Counting and Understanding
Number: Ages 4–5
© A & C BLACK**

Brothers and sisters

These children are saying how many brothers and sisters they have.

- For each pair, tick ✓ who has more .

Teachers' note This activity encourages the children to compare the size of numbers (up to 5) without counting objects. Children who find this activity difficult could use cubes to represent the number of siblings each has and then compare them directly. Alternatively, they could find and compare both numbers on a number line.

100% New Developing Mathematics
Counting and Understanding
Number: Ages 4–5
© A & C BLACK

Snail hunt

- **Read how many snails are under each pot.**
- **For each pair, colour the pot that has** `more` .

NOW TRY THIS!

- **Write a number on the pots to show**

more snails

fewer snails

Teachers' note This activity encourages the children to compare the size of numbers (up to 10) without counting objects. Children who find this activity difficult could place the correct number of cubes next to each plant pot and then compare them directly. Alternatively, they could find and compare both numbers on a number line.

**100% New Developing Mathematics
Counting and Understanding
Number: Ages 4–5**
© A & C BLACK

Football scores

- **Draw a ring around the team who scored** | more | **goals.**

| 4 | 1 |

| 3 | 5 |

| 2 | 7 |

| 8 | 6 |

NOW TRY THIS!

- **Draw a ring around the team who scored** | fewer | **goals.**

| 5 | 7 |

| 9 | 8 |

| 6 | 4 |

Teachers' note This comparing activity involves the more abstract notion of goals scored, where the number represents an action rather than an object. Demonstrate how a number line can be used to help find which number is larger in each pair.

50

**100% New Developing Mathematics
Counting and Understanding
Number: Ages 4–5
© A & C BLACK**

Baby, baby: 1

• **Colour the correct number of jelly babies.**

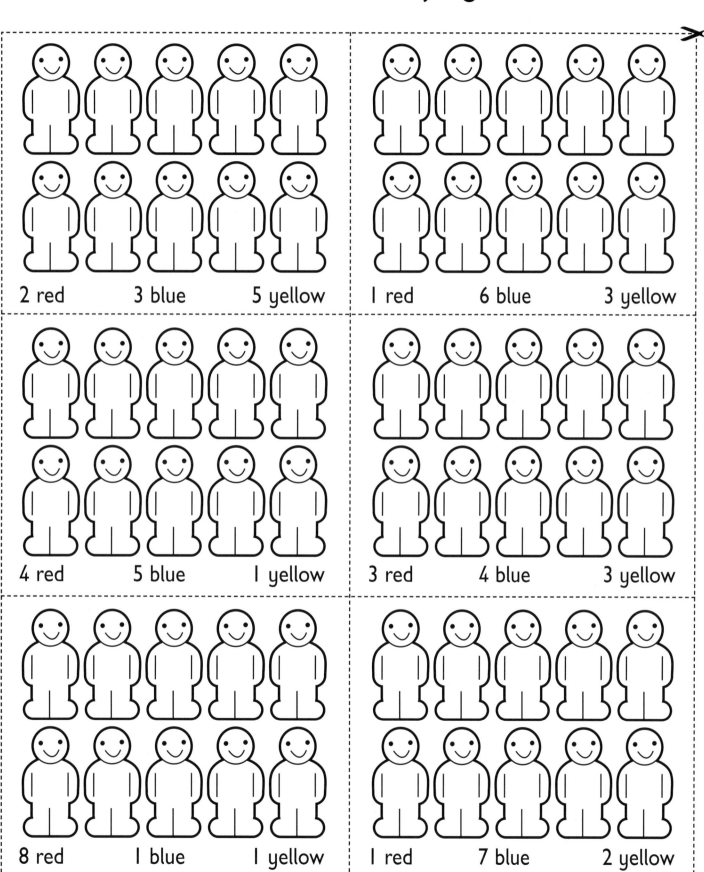

2 red   3 blue   5 yellow

1 red   6 blue   3 yellow

4 red   5 blue   1 yellow

3 red   4 blue   3 yellow

8 red   1 blue   1 yellow

1 red   7 blue   2 yellow

Teachers' note Once coloured, the cards can be used to play a game in pairs or small groups. The cards are placed face down and each child picks one. Player 1 chooses a colour (the one he/she thinks most likely to win), for example red. Players say how many reds are on their card. The player with the highest number wins the cards. Players pick a new card, player 2 chooses the colour, etc.

100% New Developing Mathematics Counting and Understanding Number: Ages 4–5 © A & C BLACK

Baby, baby: 2

• **Colour the correct number of jelly babies.**

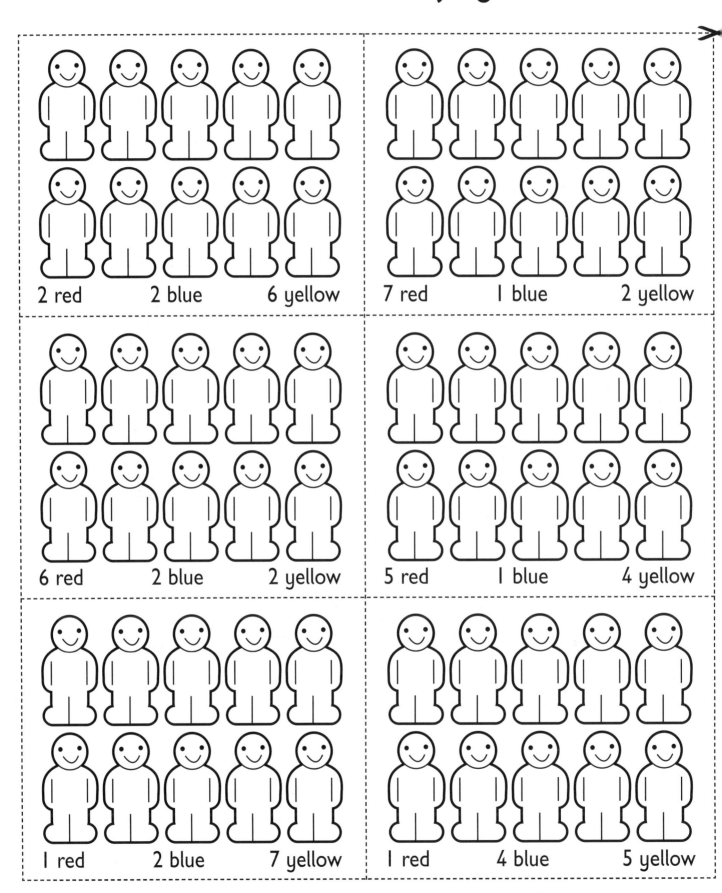

2 red   2 blue   6 yellow

7 red   1 blue   2 yellow

6 red   2 blue   2 yellow

5 red   1 blue   4 yellow

1 red   2 blue   7 yellow

1 red   4 blue   5 yellow

Teachers' note This sheet should be used in conjunction with page 51.

100% New Developing Mathematics
Counting and Understanding
Number: Ages 4–5
© A & C BLACK

Compare jars

- **Read how many sweets are in each jar.**
- **For each shelf, tick** ✓ **the jar that has** more .

- **Colour jars that have** fewer than 5 **sweets.**

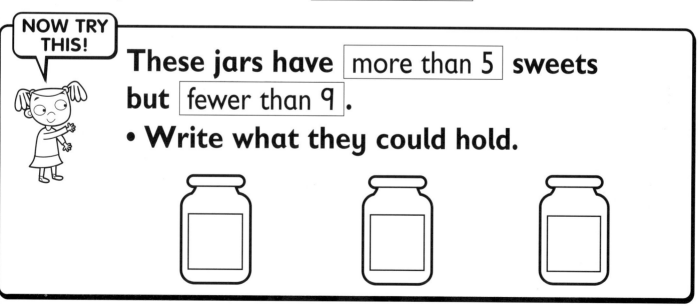

NOW TRY THIS!

These jars have more than 5 **sweets but** fewer than 9 .

- **Write what they could hold.**

Teachers' note If necessary, the children could count out cubes and make a direct comparison. Alternatively, they could find and compare both numbers on a number line. This activity could be linked with play activities in a shop. If appropriate, the children could 'pay' for each jar using 1p coins (1p for each sweet) and then compare amounts.

100% New Developing Mathematics
Counting and Understanding
Number: Ages 4–5
© A & C BLACK

More or less

These children are holding numbers.
- Which number is less or more ?

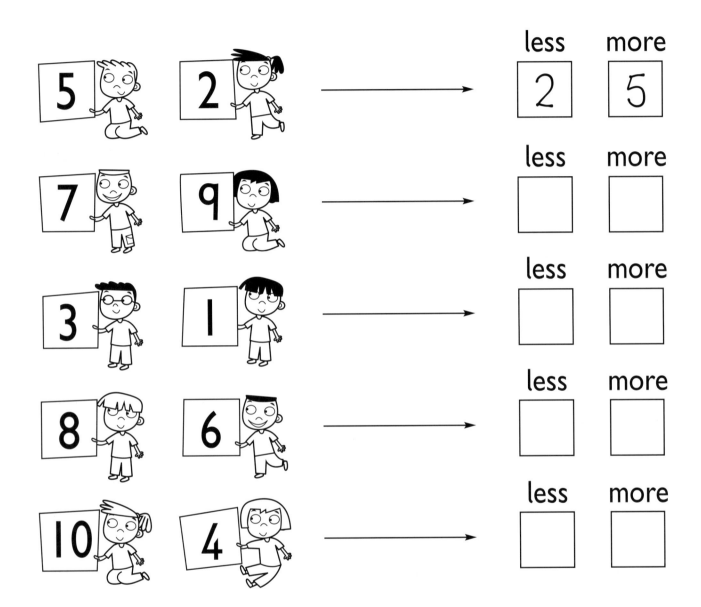

	less	more
5, 2	2	5
7, 9		
3, 1		
8, 6		
10, 4		

NOW TRY THIS!

- Write a number in each box.

less	more		less	more
3				

Teachers' note This activity uses the vocabulary 'more' and 'less' to compare two numbers. Children who are having difficulty could refer to a number line. In the extension activity, the children have to choose any number that is more than 3, and then choose a pair of numbers of their own and say which is more and which is less.

100% New Developing Mathematic
Counting and Understanding
Number: Ages 4–5
© A & C BLACK

• **Write the numbers in the correct boxes.**

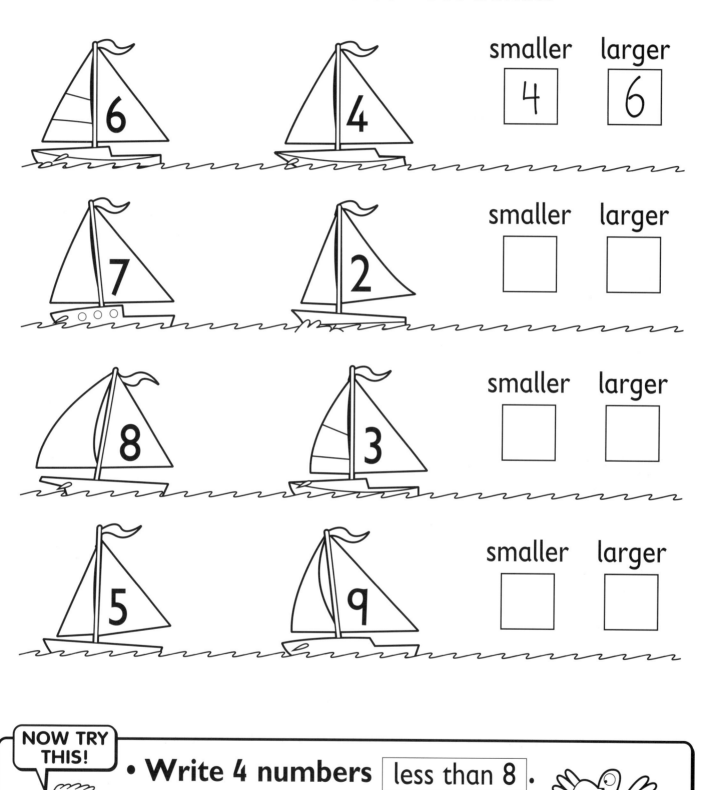

smaller larger

6   4 | 4 | | 6 |

smaller larger

7   2

smaller larger

8   3

smaller larger

5   9

NOW TRY THIS!

• **Write 4 numbers** less than 8 .

Teachers' note At the start of the lesson, give groups of children a digit card each and ask pairs to rearrange themselves in order, starting with the smallest. Use 'larger' and 'smaller' to compare the numbers.

**100% New Developing Mathematics
Counting and Understanding
Number: Ages 4–5
© A & C BLACK**

Biggest is best

- Cut out the cards.
- Stick them on the chart.

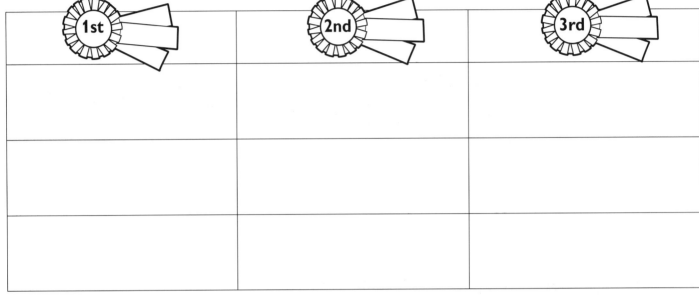

1st	2nd	3rd

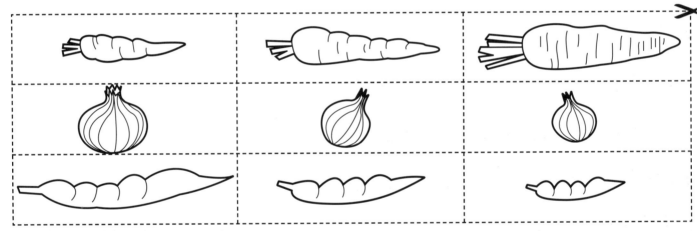

NOW TRY THIS!

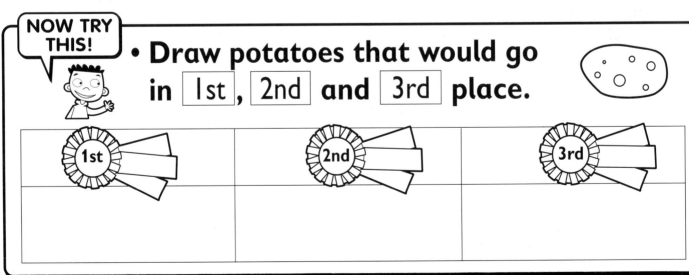

- Draw potatoes that would go in 1st , 2nd and 3rd place.

1st	2nd	3rd

Teachers' note Before beginning the activity, ensure that the children understand how prizes are given in a competition for first, second and third place. Hold up a range of different items and encourage the children to say which would come first, second and third, for example different-sized pencils, pieces of fruit or sunflowers.

100% New Developing Mathematics
Counting and Understanding
Number: Ages 4–5
© A & C BLACK

Go-kart race

- **Number the go-karts in order.**
- **Start at** $\boxed{1}$ **.**

$\boxed{1}$ $\boxed{}$ $\boxed{}$ $\boxed{}$ $\boxed{}$

- **Colour the** $\boxed{\text{3rd}}$ **go-kart** $\boxed{\text{red}}$ **.**
- **Colour the** $\boxed{\text{2nd}}$ **go-kart** $\boxed{\text{yellow}}$ **.**
- **Colour the** $\boxed{\text{1st}}$ **go-kart** $\boxed{\text{blue}}$ **.**
- **Colour the** $\boxed{\text{5th}}$ **go-kart** $\boxed{\text{green}}$ **.**
- **Colour the** $\boxed{\text{4th}}$ **go-kart** $\boxed{\text{orange}}$ **.**

NOW TRY THIS!

- **Match each go-kart to the correct box.**

$\boxed{\text{2nd}}$ $\boxed{\text{1st}}$ $\boxed{\text{5th}}$ $\boxed{\text{4th}}$ $\boxed{\text{3rd}}$ $\boxed{\text{8th}}$ $\boxed{\text{6th}}$ $\boxed{\text{7th}}$

Teachers' note During structured play activities, encourage the children to describe the positions of toy cars, using ordinal numbers. Before beginning the activity, ensure that the children know how to say the words 'first', 'second', 'third', and so on.

100% New Developing Mathematics
Counting and Understanding
Number: Ages 4–5
© A & C BLACK

Dumper trucks

These trucks are in a queue.

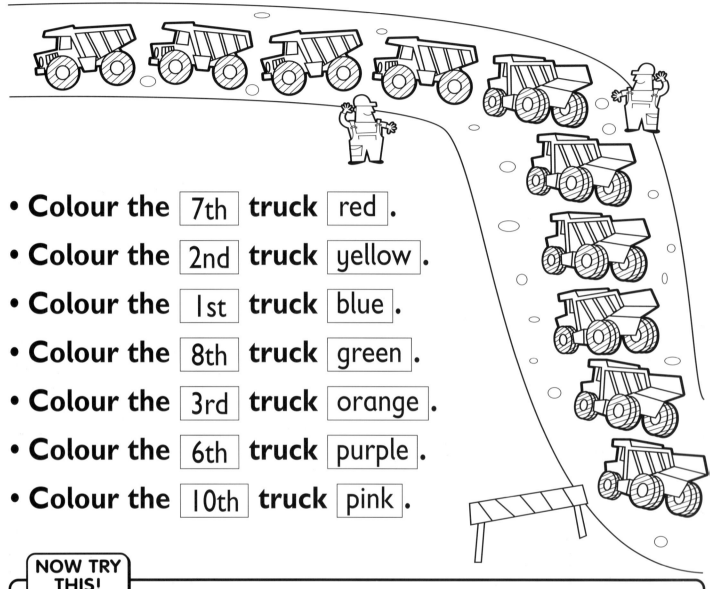

- **Colour the** [7th] **truck** [red].
- **Colour the** [2nd] **truck** [yellow].
- **Colour the** [1st] **truck** [blue].
- **Colour the** [8th] **truck** [green].
- **Colour the** [3rd] **truck** [orange].
- **Colour the** [6th] **truck** [purple].
- **Colour the** [10th] **truck** [pink].

NOW TRY THIS!

- **Match each truck to the correct box.**

| 4th | 1st | 5th | 2nd | 3rd | 9th | 6th | 10th | 7th | 8th |

Teachers' note During structured play activities, encourage the children to describe the positions of toy cars, using ordinal numbers. Before beginning the activity, ensure that the children know how to say the words 'first', 'second', 'third', and so on. Some children might find it useful to write the numbers from 1 to 10 next to the trucks.

**100% New Developing Mathematics
Counting and Understanding
Number: Ages 4–5**
© A & C BLACK

Lining-up time

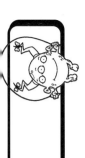

- Cut out the cards and strips.
- Glue the strips together.
- Put the cards in order under the line of children.

4th	1st	7th	2nd	3rd
9th	6th	10th	5th	8th

Emma

Joe

Li

Sam

Ruby

Tom

Nazreen

Jake

Anne

Ali

Teachers' note Invite the children to cut out the two strips and the ordinal number cards. By sticking the strips together, a queue of children is formed. The children should then put the numbers in order beneath the queue. Once the children are happy with their ordering, the strips and cards can be stuck onto a long strip of paper and displayed.

**100% New Developing Mathematics
Counting and Understanding
Number: Ages 4–5
© A & C BLACK**

Cheeky chimps

- **Work out which numbers are hidden.**
- **Write the numbers.**

- **Which numbers could this be?**

[ ] or [ ] or [ ]

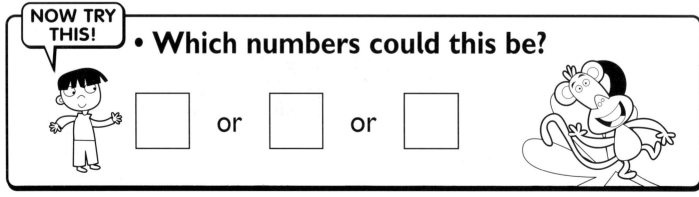

Teachers' note At the start of the lesson, encourage the children to draw numerals in the air and to describe the shapes as they draw them. For example: '3 is half round, half round'; '7 is across and down'. Ask the children to watch as you draw a numeral in the air, focusing on whether the sides are curved or straight.

100% New Developing Mathematics Counting and Understanding Number: Ages 4–5
© A & C BLACK

Find the numbers

• **Find the numbers in this picture and colour them.**

NOW TRY THIS!

• **Draw your own picture with hidden numbers.**

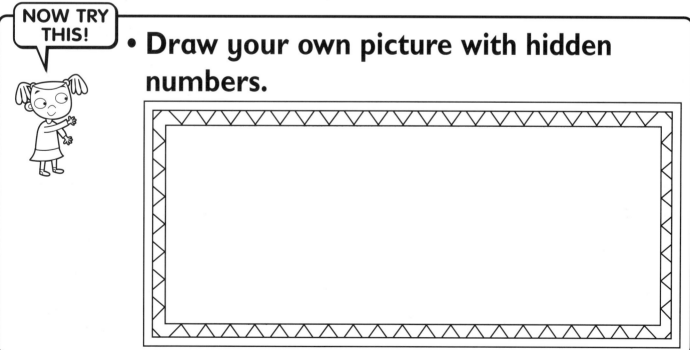

Teachers' note Here children distinguish numerals from other letters and shapes. If a child colours the letter S (in the form of a snake), discuss how the number 5 is usually written with a line going straight across the top and that the starting position for writing the number 5 is top left, rather than top right. Encourage the children to draw the number 5 in the air to reinforce this.

**100% New Developing Mathematics
Counting and Understanding
Number: Ages 4–5
© A & C BLACK**

Tracing tracks

- ## Trace over the numbers. **Start on the dot.**

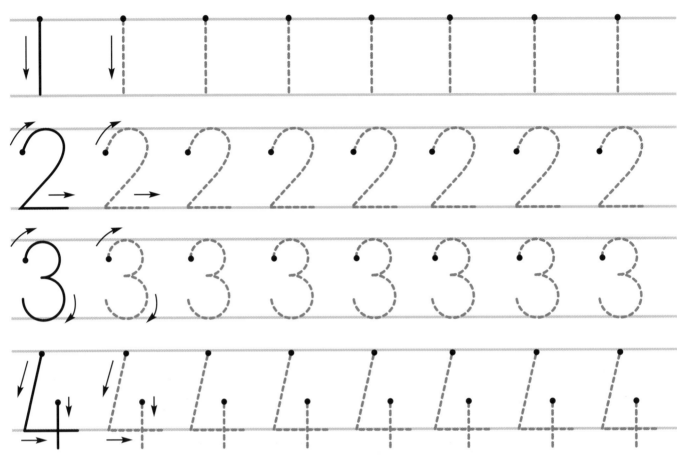

NOW TRY THIS!

- ## Write the numbers.

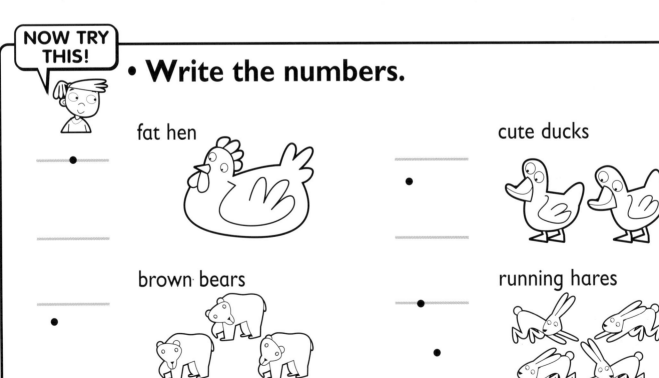

fat hen

cute ducks

brown bears

running hares

Teachers' note It is important that the children appreciate the meaning of the numerals they are writing. Begin the lesson by inviting nhe children to find sets of 1, 2, 3 or 4 objects around the room, for example 2 doors, 4 windows, 1 teacher. Ask the children to practise drawing the numerals in the air, into the palm of the other hand, on the back of the child in front, etc.

**100% New Developing Mathematics
Counting and Understanding
Number: Ages 4–5**
© A & C BLACK

Tracing tractors

• **Trace over the numbers. Start on the dot.**

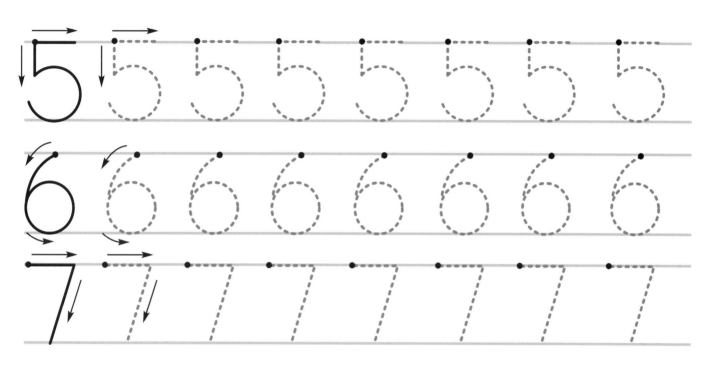

NOW TRY THIS!

• **How many tractors in each set?**
• **Write the number four times.**

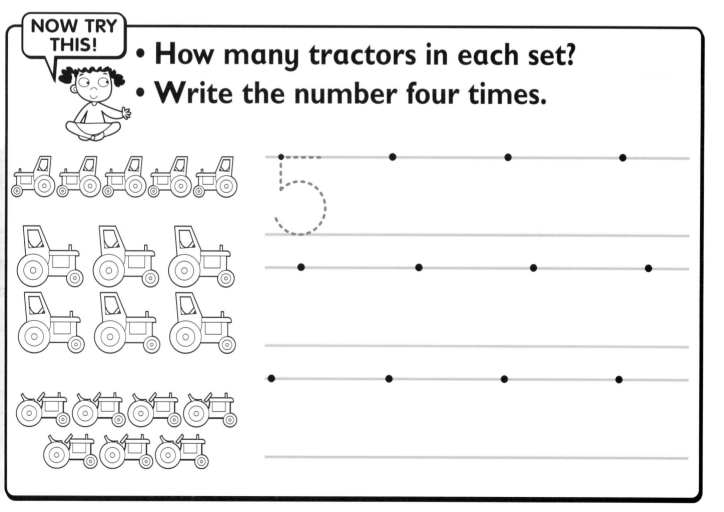

Teachers' note It is important that the children appreciate the meaning of the numerals they are writing. Begin the lesson by inviting the children to find sets of 5, 6, or 7 objects around the room, for example 5 tables, 6 pencils, 7 chairs. Ask the children to practise drawing the numerals in the air, into the palm of the other hand, on the back of the child in front, etc.

**100% New Developing Mathematics
Counting and Understanding
Number: Ages 4–5
© A & C BLACK**

Tracing trainers

• **Trace over the numbers. Start on the dot.**

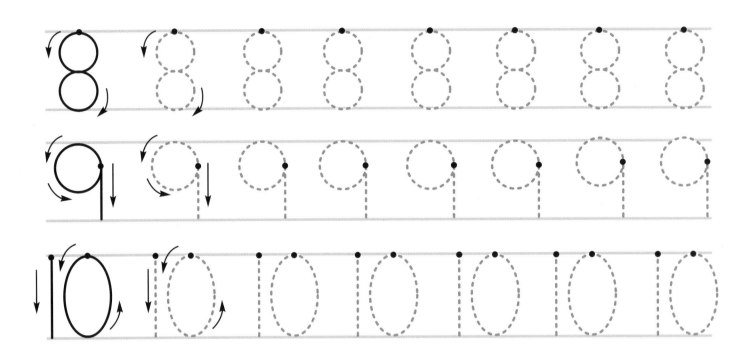

NOW TRY THIS!

• **How many trainers in each set?**
• **Write the number four times.**

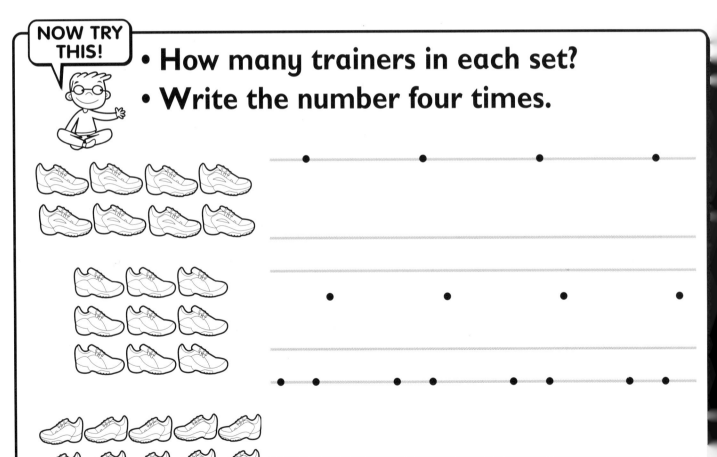

Teachers' note It is important that the children appreciate the meaning of the numerals they are writing. Begin the lesson by inviting the children to hold up 8, 9 or 10 fingers, or to collect 8, 9 or 10 counters or cubes from, for example, a drawer. Ask the children to practise drawing the numerals in the air, into the palm of the other hand, on the back of the child in front, etc.

100% New Developing Mathematics
Counting and Understanding
Number: Ages 4–5
© A & C BLACK